MES AMIS LES ANIMAUX DU ZOO

LES PINGOUINS

AMY CULLIFORD

Un livre de la collection
Les racines de Crabtree

CRABTREE
Publishing Company
www.crabtreebooks.com

Soutien de l'école à la maison pour les parents, les gardiens et les enseignants

Ce livre aide les enfants à se développer grâce à la pratique de la lecture. Voici quelques exemples de questions pour aider le lecteur ou la lectrice à développer ses capacités de compréhension. Les suggestions de réponses sont indiquées en rouge.

Avant la lecture

- De quoi ce livre parle-t-il?
    - *Je pense que ce livre parle des pingouins.*
    - *Je pense que ce livre parle des pingouins dans les zoos.*

- Qu'est-ce que je veux apprendre sur ce sujet?
    - *Je veux apprendre à quoi ressemblent les pingouins.*
    - *Je veux apprendre ce que les pingouins aiment faire.*

Pendant la lecture

- Je me demande pourquoi...
    - *Je me demande pourquoi les pingouins ont des ailes.*
    - *Je me demande pourquoi les pingouins aiment le froid.*

- Qu'est-ce que j'ai appris jusqu'à présent?
    - *J'ai appris que certains pingouins peuvent sauter.*
    - *J'ai appris que la plupart des pingouins sont noir et blanc.*

Après la lecture

- Nomme quelques détails que tu as retenus.
    - *J'ai appris que certains pingouins aiment nager.*
    - *J'ai appris que les pingouins pondent des oeufs.*

- Lis le livre à nouveau et cherche les mots de vocabulaire.
    - *Je vois le mot **pingouin** à la page 3 et le mot **neige** à la page 6. L'autre mot de vocabulaire se trouve à la page 14.*

Voici un **pingouin**.

La plupart des pingouins sont noir et blanc.

La plupart des pingouins aiment la **neige**.

Certains pingouins peuvent sauter.

Certains pingouins pondent des **oeufs**.

Tous les pingouins aiment l’eau.

Liste de mots

Mots courants

aiment
blanc
certains
et
la
noir
peuvent
plupart
sont
tous
un
voici

La boîte à mots

neige

oeufs

pingouin

32 mots

Voici un **pingouin**.

La plupart des pingouins sont noir et blanc.

La plupart des pingouins aiment la **neige**.

Certains pingouins peuvent sauter.

Certains pingouins pondent des **oeufs**.

Tous les pingouins aiment l'eau.

CRABTREE
Publishing Company

MES AMIS LES ANIMAUX DU ZOO

LES PINGOUINS

Autrice : Amy Culliford
Conception : Rhea Wallace
Développement de la série :
James Earley
Correctrice : Janine Deschenes
Conseils pédagogiques :
Marie Lemke M.Ed.
Traduction : Annie Evearts
Coordinatrice à l'impression :
Katherine Berti
Références photographiques :
Shutterstock : slowmotiongli : couverture; Gary Cox : p. 1; Gary Cox : p. 3, 14; M Kunz : p. 5; Jkornwika Supisa : p. 7, 14; nwdph : p. 8-9; Roger Clark : p. 11, 14; David Herraez Calzada : p. 13

Crabtree Publishing Company
www.crabtreebooks.com 1-800-387-7650

Publié aux États-Unis
Crabtree Publishing
347 Fifth Avenue
Suite 1402-145
New York, NY, 10016

Publié au Canada
Crabtree Publishing
616 Welland Ave.
St. Catharines, Ontario
L2M 5V6

Imprimé au Canada/062021/CPC

Catalogage avant publication de Bibliothèque et Archives Canada
Titre: Les pingouins / Amy Culliford ; texte français d'Annie Evearts.
Autres titres: Penguins. Français.
Noms: Culliford, Amy, 1992- auteur.
Description: Mention de collection: Mes amis les animaux du zoo | Les racines de Crabtree | Traduction de : Penguins. | Comprend un index.
Identifiants: Canadiana (livre imprimé) 20210272562 | Canadiana (livre numérique) 20210272589 | ISBN 9781039607620 (couverture souple) | ISBN 9781039607682 (HTML) | ISBN 9781039607743 (EPUB) | ISBN 9781039607804 (livre numérique avec narration)
Vedettes-matière: RVM: Manchots (Oiseaux)—Ouvrages pour la jeunesse. | RVMGF: Documents pour la jeunesse.
Classification: LCC QL696.S473 C8514 2022 | CDD j598.47—dc23

GERMS AND DISEASE

WHAT ARE BACTERIA?

By Kathryn Kroe

Cavendish Square
New York

Published in 2023 by Cavendish Square Publishing, LLC
29 East 21st Street, New York, NY 10010

Website: cavendishsq.com

Disclaimer: Portions of this work were originally authored by Margaux Baum and Lesli J. Favor and published as *Bacteria*. All new material this edition authored by Kathryn Kroe.

This publication represents the opinions and views of the author based on their personal experience, knowledge, and research. The information in this book serves as a general guide only. The author and publisher have used their best efforts in preparing this book and disclaim liability rising directly or indirectly from the use and application of this book.

All websites were available and accurate when this book was sent to press.

Library of Congress Cataloging-in-Publication Data

Names: Kroe, Kathryn, author.
Title: What are bacteria? / Kathryn Kroe.
Description: New York : Cavendish Square Publishing, [2023] | Series: Germs and disease | Includes index.
Identifiers: LCCN 2021058067 | ISBN 9781502665409 (library binding) | ISBN 9781502665393 (paperback) | ISBN 9781502665416 (ebook)
Subjects: LCSH: Bacteria—Juvenile literature.
Classification: LCC QR74.8 .K76 2023 | DDC 579.3—dc23/eng/20211221
LC record available at https://lccn.loc.gov/2021058067

Editor: Kate Kroe
Copyeditor: Jane Katirgis
Designer: Andrea Davison-Bartolotta

The photographs in this book are used by permission and through the courtesy of: Cover Jarun Ontakrai/Shutterstock.com; series art (backgrounds) CHIARI VFX/Shutterstock.com; p. 5 Amorn Suriyan/Shutterstock.com; p. 6 fusebulb/Shutterstock.com; p. 8 Aldona Griskevicene/Shutterstock.com; p. 9 Olga Bolbot/Shutterstock.com; p. 10 vchal/Shutterstock.com; p. 12 Rudy Balasko/Shutterstock.com; p. 13 Desizned/Shutterstock.com; p. 14 Pixelbender36/Shutterstock.com; p. 16 Far700/Shutterstock.com; p. 18 Andrey_Popov/Shutterstock.com; p. 20 Handmade Pictures/Shutterstock.com; p. 21 Zigzag Mountain Art/Shutterstock.com; p. 22 WAYHOME studio/Shutterstock.com; p. 23 Pixel-Shot/Shutterstock.com; p. 24 TojePhoto/Shutterstock.com; p. 26 Daisy Daisy/Shutterstock.com; p. 28 Marina Demeshko/Shutterstock.com; p. 30 The Toidi/Shutterstock.com; p. 31 Tatjana Baibakova/Shutterstock.com; p. 33 Kateryna Kon/Shutterstock.com; p. 34 Ana Hollan/Shutterstock.com; p. 36 B Brown/Shutterstock.com; p. 39 jukurae/Shutterstock.com; p. 40 Elena Hramova/Shutterstock.com.

Some of the images in this book illustrate individuals who are models. The depictions do not imply actual situations or events.

CPSIA compliance information: Batch #CSCSQ23: For further information contact Cavendish Square Publishing LLC, New York, New York, at 1-877-980-4450.

Printed in the United States of America

CONTENTS

INTRODUCTION

We can't see bacteria, but they are everywhere—in the air, in the soil, on our hands, and in our bodies. These microscopic, single-celled organisms have inhabited Earth far longer than humankind and live in habitats all over the planet. The study of bacteria provides fascinating insight into early history and continually evolves as we discover new bacteria. While we first observed bacteria in the 1600s, we didn't even begin to truly understand them until the 1860s.

Throughout human history, bacteria have been the source of some of the deadliest diseases and epidemics. Bubonic plague, caused by a bacterium, was devastating in the 14th century. Commonly known as the Black Death, the highly contagious disease wiped out huge portions of the European population. At that time, with no true understanding of germs, people did not know how to protect themselves from this bacterium. Leprosy, tuberculosis, and cholera are other bacterial diseases that have had a huge impact on human populations and health throughout the ages.

So, should we fear bacteria? While some bacteria are certainly harmful, others are crucial to our health and a vital part of life. You may be surprised to find out some of the everyday places we can find healthy bacteria. While we cannot avoid bacteria, we can certainly better understand them.

People are now able to study bacteria using a microscope to observe how they look and behave.

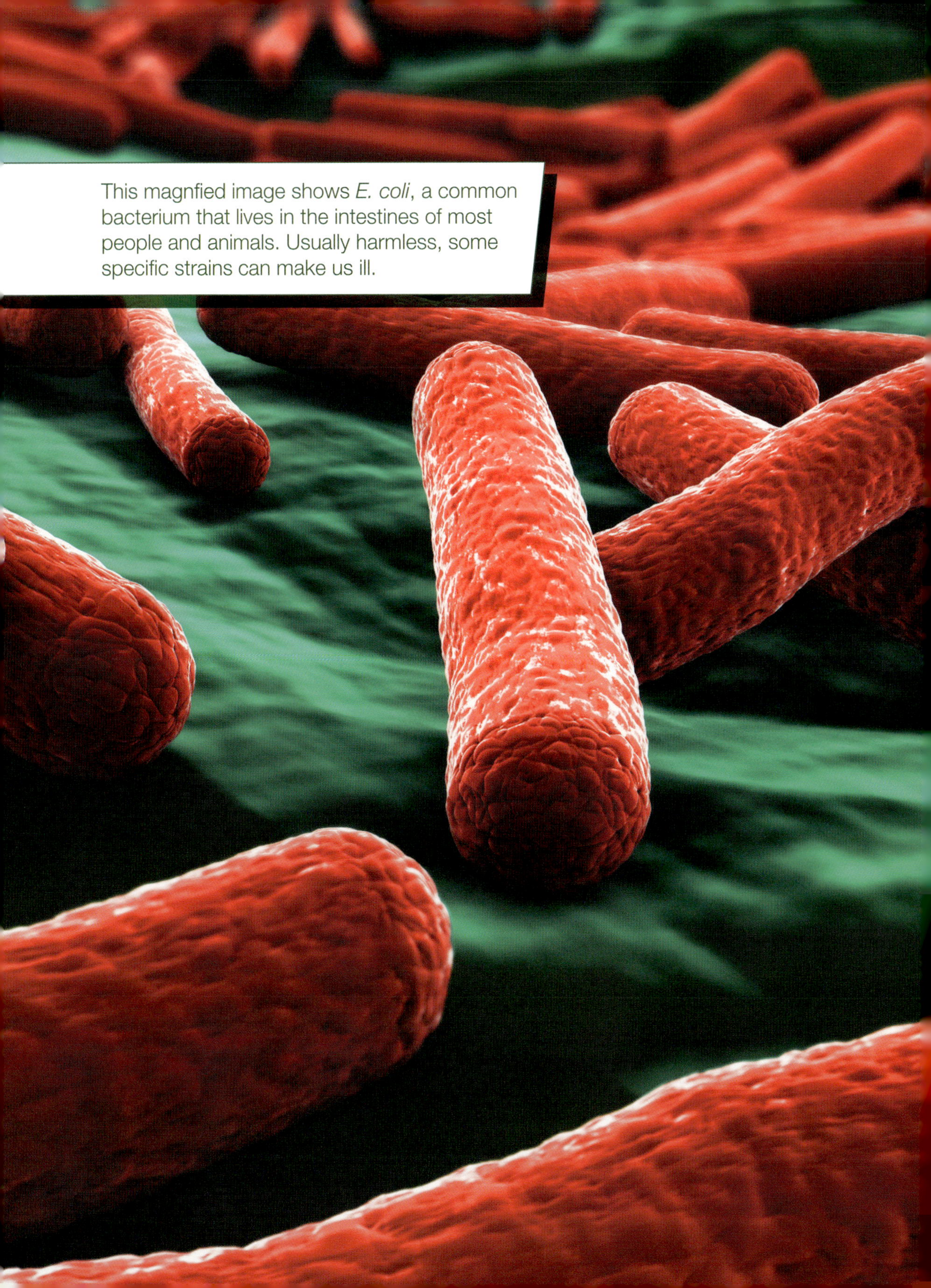

This magnfied image shows *E. coli*, a common bacterium that lives in the intestines of most people and animals. Usually harmless, some specific strains can make us ill.

CHAPTER 1

WHAT ARE THE CHARACTERISTICS OF BACTERIA?

Scientists classify living things into biological kingdoms. Humans belong to the animal kingdom, and most bacteria belong to the bacteria kingdom. The organisms in this kingdom are single-celled but still complex. The bacteria kingdom includes all bacteria except archaea, which live in extreme environments; these bacteria are genetically distinct from the bacteria we are most familiar with and thus classified in a different kingdom, archaea.

BACTERIA CELL STRUCTURE

Each bacterium is formed of only one cell, which has a very simple structure. Many kinds of cells have a cell wall surrounding a cell's cytoplasm, and inside the cytoplasm are organelles and a nucleus containing genetic material. Bacteria lack this complex internal structure. All cellular matter floats freely within the cell. A few kinds of bacteria even lack cell walls.

Bacteria are **prokaryotes**. This means the bacteria do not have a distinct nucleus. Regular cells with nuclei and other organelles such as mitochondria are called eukaryotes and are considered a later evolutionary development. In prokaryotes, rather than being contained in a nucleus, DNA floats freely within the cell.

PROKARYOTIC VS. EUKARYOTIC CELLS

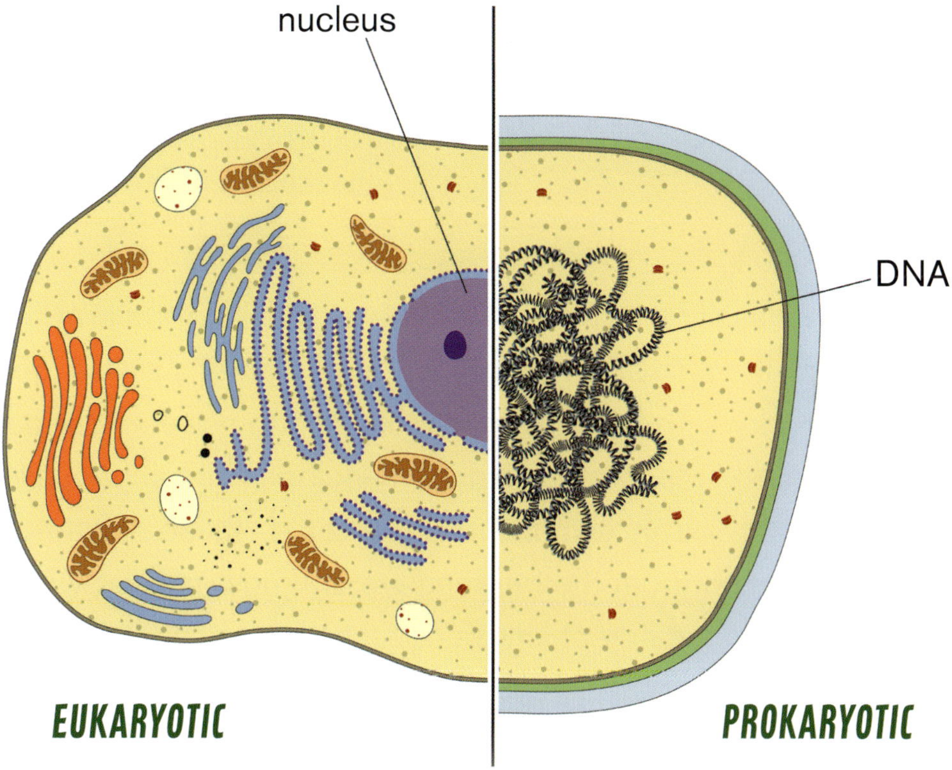

Bacteria are prokaryotes. Notice how the DNA is not contained in a nucleus, unlike in the eukaryotic cell.

Shape and Movement

Bacteria are grouped into three types of cell shapes: spherical, rodlike, and spiral. Coccus cells are spherical. Certain types of cocci (the plural of coccus) cause strep throat and pneumonia, while other types live harmlessly within humans. Bacillus cells are rodlike, cylindrical tubes. Bacilli can cause tuberculosis and anthrax, among other diseases. Spirillum cells have a loose spiral shape. Also called spirochetes, they can cause Lyme disease and other infections.

SHAPES OF BACTERIA

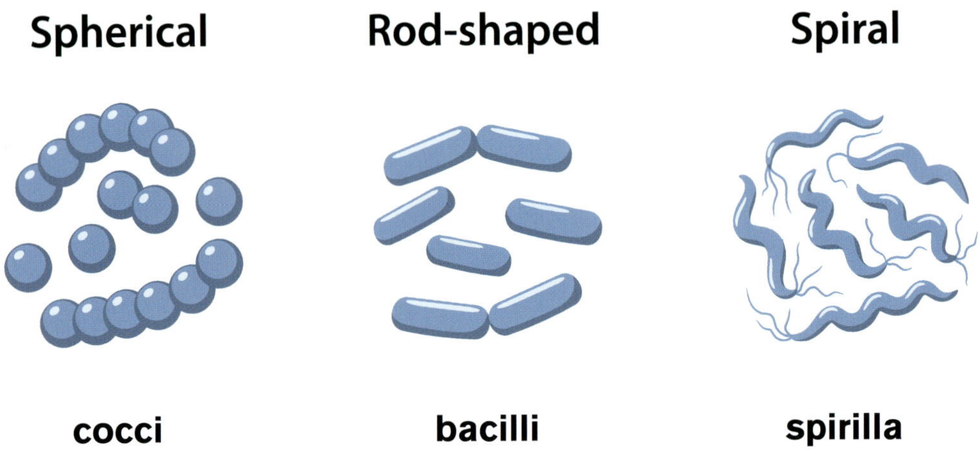

Bacteria come in three basic shapes, as seen above.

Bacteria are single-celled, but sometimes bacteria stick together as pairs, clusters, or lines of cells. When clinging together as a pair, they are called diplo. For example, a stuck-together pair of coccus bacteria is called diplococcus. *Diplococcus pneumoniae* is one cause of conjunctivitis, an eye inflammation. When clustering together, they are called staphylo. Staphylococcus bacteria cause staph infections, meningitis, food poisoning, and other ailments. When linked in chains, they are called strepto. Streptococci, for example, are chains of coccus bacteria. Looking much like strings of beads, they cause strep throat, scarlet fever, sinusitis, and other infections.

Some bacteria move themselves, but others are passengers of the substance in which they live. The mobile ones vary in their capabilities. A slimy bacterium can slide slowly using its own

slippery coating. Those with **flagella** propel themselves forward in a swimming motion. Spirochetes use structures similar to flagella, located beneath the surface of the cell membrane, to swim. When sensing danger such as a toxic chemical, bacteria evade it. When sensing food, hungry bacteria approach it.

Other bacteria, though immobile on their own, may float through the air on a speck of dust or on an insect, hurl through the air in the droplets of a sneeze, or ride along on the skin of an animal.

Reproductive Methods

Bacteria are present all over Earth in such great numbers because of their ability to reproduce rapidly. In a process called binary **fission**, a bacterium splits in two. Each half then splits in two, and those four parts each split in two. The cell division continues, limited only by available space and nutrients. This is called asexual reproduction. The original cell is the mother cell, and the two resulting ones are daughter cells. Originally half the mother's size, the daughters grow. Bacterial life spans vary, ranging from minutes to hours, but they are all very short.

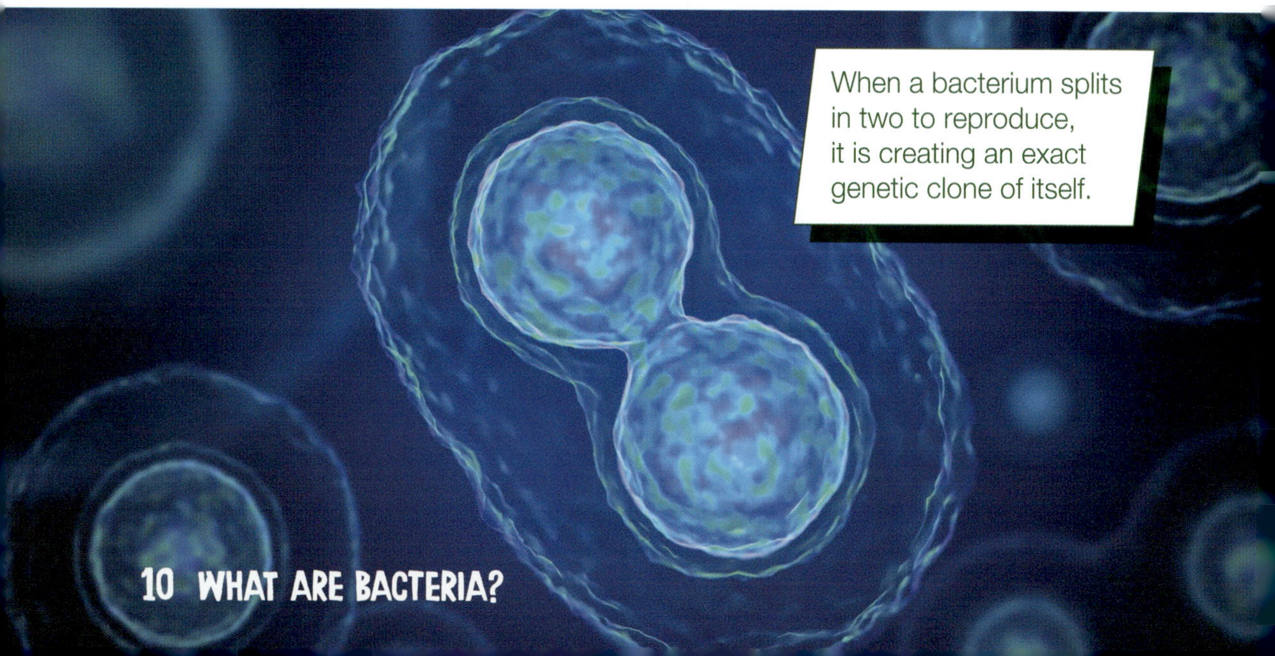

When a bacterium splits in two to reproduce, it is creating an exact genetic clone of itself.

In favorable conditions, one bacterium can divide as often as once every 20 minutes. The tip of a finger might hold 10,000 bacteria at noon. If they divide every 20 minutes, 1 hour later there would be 80,000! If humans multiplied on this scale, a family of 4 would be a family of 32 in 1 hour! More commonly, bacteria divide once every 2 or 3 hours and others only once every 16 hours.

While most bacteria reproduce through binary fission, some reproduce through **conjugation**. In conjugation, two bacteria join and exchange DNA before pulling back apart. While cells produced by fission are identical to the mother—they are clones—those produced by conjugation share DNA from both parents. They are unique.

Occasionally, a mutant bacterium arises. It has DNA slightly different from the parent cells' DNA. Consequently, the mutant may survive conditions that would kill its parents. Researchers believe that mutation is responsible for bacteria's ability to adapt to such a wide array of living conditions, both now and throughout the **microbes**' long history.

Bacteria Everywhere

Bacteria live everywhere—in more environments and habitats than any other kind of living thing. Salt water, fresh water, sand, dirt, animals, humans, plants, air, rotting flesh—these are all desirable habitats for bacteria of all kinds. Scientists studying rocks and dirt in mines 1,300 feet (390 meters) deep inside Earth found bacteria there. Other scientists studying air 19 miles (30 kilometers) above Earth's surface found bacteria there too. Some bacteria survive freezing or boiling that would kill other living things. Some types, called **anaerobes**, thrive in places with no oxygen. They

obtain oxygen from their food. Others, called aerobes, rely on oxygen-rich air for survival, much like humans and animals. Still others can survive either with or without air.

A few kinds of bacteria form spores. A spore is a resting stage, making it possible for the cell to survive adverse conditions. Some spores survive in boiling water for hours, while others survive chemical poisons. Other types of spores endure extreme cold or dehydration. With their tough, dry coats, some spores have survived for hundreds of years. The ruins of ancient Rome have yielded spores nearly 2,000 years old. Other spores have lasted longer. Off the coast of California, in the sediment of the Pacific Ocean, spores were found from 5,800 years ago. In Minnesota's Elk Lake, sediment yielded spores believed to be the oldest ever found, at around 7,000 years old.

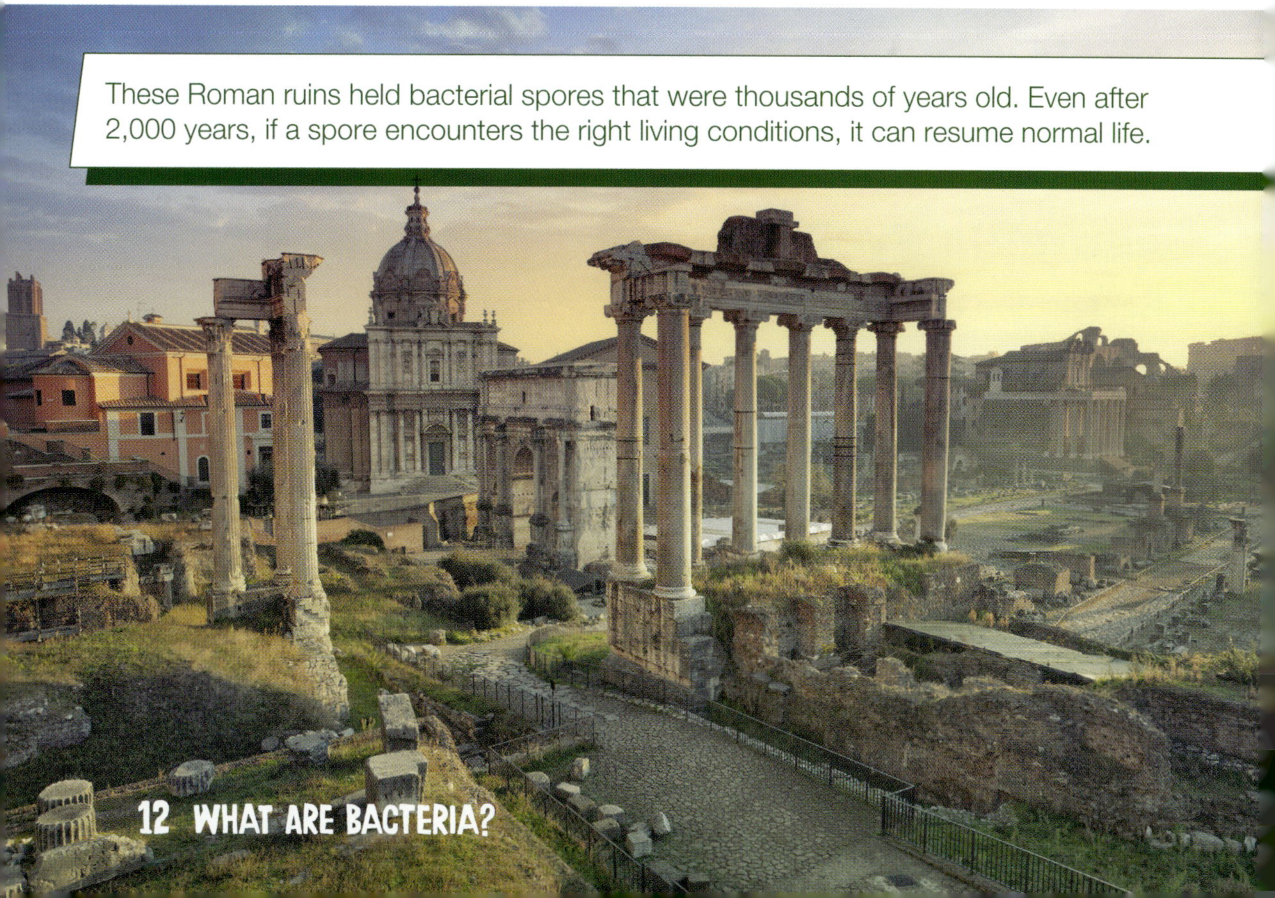

These Roman ruins held bacterial spores that were thousands of years old. Even after 2,000 years, if a spore encounters the right living conditions, it can resume normal life.

When a spore encounters favorable living conditions once again, the tough coat gives way to a healthy cell. The bacterium resumes normal, active life. Moisture, warmth, and nutrients are conditions favorable to reactivating a dormant bacterium.

Certain bacteria live in the rumen, a stomach section in some animals. Here they help break down cellulose, the stiff matter in plants, during digestion. Without the aid of these bacteria, cows, sheep, goats, and other animals would receive little nutrition from the green plants they graze on. Humans, too, rely on bacteria to help digest food. Colonies of *Escherichia coli* (*E. coli*) thrive in the intestines, helping to break down food and release nutrients in usable form for the human body.

The human mouth is a bacterium's paradise. Warm and moist, it provides an ideal habitat for billions of bacteria. Sugars and starches in food pass through the mouth, feeding the bacteria as well as their host. Colonies of the microbes live on the tongue, teeth, and gums, in crevices between teeth, and in dental plaque.

Our mouth provides a cozy home for bacteria. That is why regular dental care is so important to our health.

Needed Nutrients

Bacteria need nutrients to survive, just as people do. The bacterium takes in food through the cell wall or cell membrane, which separates the insides of a cell from the outside world. First, the bacterium secretes **enzymes** onto food outside. The enzymes break down the food into soluble form. Being soluble means the food can dissolve in a liquid, and in this form, the bacterium absorbs the food. Inside the cell, enzymes further process the food.

For some bacteria, mineral compounds provide the best nutrients. Using sulfur, minerals in water, and carbon dioxide from air, these bacteria manufacture carbohydrates, fats, proteins, and vitamins. Because they create their own nourishment, they are called autotrophic, or self-nourishing. Other bacteria need more complex foods. Parasitic bacteria depend on a living host to provide nourishment. Bacteria called saprophytes feast on rotting animal **carcasses**, dead bugs, and other types of decaying flesh.

This dead bug provides a delicious feast for certain types of bacteria.

Just as plants use photosynthesis to manufacture nutrients, so do some anaerobic bacteria. These bacteria contain substances that capture energy from sunlight and use it to make food. Whereas plants produce oxygen as a waste product, anaerobic bacteria do not. However, cyanobacteria also employ photosynthesis, and they do give off oxygen as waste.

After digestion, the bacterium expels waste matter through the cell wall or membrane. The waste products of bacteria that are poisonous to other living things are called toxins.

Bacteria's Role

One of bacteria's valuable roles in nature is to assist in the decay of dead organisms. Along with earthworms and fungi, bacteria break down **organic** matter such as fallen leaves, dead plants and wood, and animal carcasses. Bacteria work during the final stages of decomposition, after earthworms and fungi have broken the matter into tiny pieces. As they feast on organic molecules, bacteria break them down into basic elements, including nitrates, minerals, water, and carbon dioxide. They release these elements back into the soil and air, where they originated. Thus enriched, the soil and atmosphere can support a new cycle of life. If it weren't for decomposition from bacteria, dead organic matter would pile up, tying up the basic elements needed for new life.

Another process involves breaking down solid waste from humans and animals. In breaking down manure, for example, bacteria release nitrogen, phosphorus, and potassium, which are important nutrients in fertile soil. For this reason, decaying manure is used as soil fertilizer. Billions of pounds of waste are produced daily

worldwide. Without the work of bacteria, aided by other microbes including fungi and protozoa, waste would make Earth unlivable.

The bacterium's role as an agent of decay is crucial in maintaining the natural balance of life itself. Bacterial processes help introduce valuable nutrients into the soil and carbon dioxide into the air, thus benefiting plant life and, by extension, animal life too.

For something so small, bacteria are surprisingly complex. Their existence is vital to our existence, both personally and on a larger scale. They are crucial in life-or-death matters for human populations around the world, as well as for plants and animals too. Still, bacteria don't always help us, and in some cases, they can cause serious harm.

Because of bacteria, manure becomes a valuable soil additive to enhance plant growth.

BACTERIA BREATH

Have you ever woken up in the morning and realized your breath didn't smell very fresh? Your mouth is a great home for bacteria and other microorganisms that can cause bad breath. In fact, the mouth is home to 700 species of microbes, good and bad. Warmth, moisture, and sugars from the foods we eat make it a welcome home for germs. Unfortunately, that means they also leave some unwelcome side effects.

While we can't avoid bacteria in our mouths, we can take steps to manage and reduce them. Regular brushing and flossing remove bacteria from our teeth and gums. Eating fiber-rich fruits and vegetables and certain other foods also helps keep bacteria at bay. Sometimes an antibacterial mouthwash helps too.

Why is bad breath often worse in the morning? Our saliva helps clear out bad bacteria. While we're sleeping, saliva production is significantly reduced, and mouth bacteria has a better chance to breed. Brushing our teeth in the morning helps clean those bad bacteria out, leaving our breath fresh and clean!

WHAT DO YOU THINK?

1. Why are bacteria able to live in so many different habitats?
2. If bacterial life spans are short, how do bacteria continue to be so numerous?
3. What do you think would happen if bacteria didn't assist in decomposition?

The refrigerator is a modern convenience that keeps us healthy. Before refrigeration became common, food was much harder to keep safe from bacteria.

CHAPTER 2

WHEN ARE BACTERIA HARMFUL?

While bacteria are often helpful, their functions aren't specifically for helping or harming other organisms. Bacteria go through their life cycle, and sometimes the toxins they release make us sick. Certain bacterial germs can just make us feel unwell, while others can be fatal if they go undetected and untreated.

FOODBORNE BACTERIA

One reason many foods need to be washed and prepared carefully is because of the prevalence of bacteria on or in them. Improper food preparation, including undercooking, can lead to varying degrees of sickness, typically called food poisoning. Some bacteria make food inedible or harmful because the food is spoiled.

Bacteria active in the decay process are responsible for food spoilage. Moderate temperatures and some moisture are ideal for bacteria to thrive in many types of food. However, refrigeration helps slow spoilage, and freezing can halt spoilage, at least until the food is thawed. At this time, the bacteria become active again. Heat used in cooking and baking then kills many types of bacteria. While spoiled food may contain bacteria, it is often harmless and not infectious. Still, foodborne bacteria can be pathogenic, or cause disease, so proper food safety is crucial.

In certain foods, bacteria bring about spoilage more readily than in others. A cheesy slice of pizza left sitting on a kitchen counter will spoil much faster than a piece of dry toast. A glass of milk left out on the counter will sour more quickly than a glass of sugary, acidic lemonade. A can of salt, a box of sugar cubes, a bag of flour, and a box of dry breakfast cereal are all examples of foods that do not readily support bacterial growth. The fact that these products are dry goods helps keep them relatively safe.

On the other hand, meats, cheeses, and prepared foods are hospitable to bacterial growth, due in part to their moisture content. In general, most types of bacteria do not grow in dry or salty foods. For this reason, drying and salting are effective methods of food preservation. Similarly, very sugary foods are not conducive to bacterial growth, nor are acidic foods such as vinegar.

Meat would not usually be considered a shelf stable product because of the risk of bacteria. However, in a product such as beef jerky, removing the moisture and adding salt makes it far less likely that bacteria will survive on it.

Botulin, a poisonous bacterial waste product, causes botulism. Even in tiny amounts, the toxin attacks the human nervous system and can be fatal. Botulin comes from a spore that survives boiling water for up to five hours. Home-canned foods are most susceptible to botulin. Superheated with steam, commercially canned foods are less likely to carry the toxin. Whatever its origin, an unopened jar whose suction seal has popped up or a swollen can is likely to contain botulin.

While canning can be a great way to use and store foods at the end of a season, it is important to take extra care to prepare them correctly to prevent the growth of botulin.

Disease-Causing Bacteria

While some bacterial infections are foodborne, others result when bacteria invade the body by other means. These bacteria are responsible for infections ranging from acne to pink eye to pneumonia. Some of these pathogens multiply in the body so rapidly that they interfere with the functions of tissues and organs. Others excrete toxins, chemicals that are poisonous to the human body.

Pneumonia is a respiratory infection resulting from bacteria that multiply rapidly in the lungs. As the lungs become filled with bacteria, the person has increasing difficulty breathing. Eventually, the victim can suffocate and die.

Hemophilus bacteria are mostly harmless, except for *Hemophilus influenzae*, which causes whooping cough and meningitis. Whooping cough, characterized by coughing and vomiting, is highly contagious. Meningitis is an inflammation of the membranes covering the brain and spinal cord. Meningitis is also caused by other bacteria, including meningococcus and various kinds of

Whooping cough is a serious bacterial disease, especially for children. While there is a vaccine to prevent it, it cannot be given until an infant is two months old. Many deaths caused by this disease are in newborn infants who have not yet been vaccinated. Parents and adults who are going to be around an infant can get a booster of the vaccine to help protect both themselves and the newborn.

pneumococci, streptococci, and staphylococci. Most frequently, children under the age of 10 suffer from the disease. Early diagnosis is vital because brain damage and death may result if left untreated.

A soil bacterium, *Clostridium tetani*, lurks in spore form, entering scratches or puncture wounds that are contaminated with dirt. Once in the wound, the spore is activated by rotting tissue. It releases a toxin that enters the nervous system and spinal cord, causing tetanus. Also known as lockjaw, a tetanus infection causes severe, uncontrollable muscle contractions. Spasms of the jaw muscles give the disease its nickname. The disease can be fatal, though it is preventable with a vaccine.

The list of bacterial infections goes on. Streptococcus causes scarlet fever. *Mycobacterium bovis* causes tuberculosis. Carried by ticks, *Borrelia burgdorferi* causes Lyme disease. Rickettsia causes typhus and spotted fever. The *Chlamydia trachomatis* bacterium causes chlamydia, a sexually transmitted disease that can result in the inability to have children. Leprosy, more common in centuries past but still present today, results from *Mycobacterium leprae*. *Legionella pneumophila* bacteria cause Legionnaire's disease. Found in refrigerated foods such as potato salad and cold cuts, the bacterium *Listeria monocytogenes* causes serious food poisoning.

Not all pathogens and toxins are life threatening. For instance, acne is a

While acne can be painful in some cases, for most people it is not a serious infection. It is caused by bacteria and oil clogging the pores of the skin.

LEGIONNAIRE'S DISEASE

On July 21, 1976, 4,000 American Legion members gathered in Philadelphia, Pennsylvania, for an annual convention. The members spent four days together before returning home at the end of the gathering. By August 2, twelve of those members had died and three dozen more had been hospitalized. Because these men had all been together, fear that the illness they were suffering from was viral, like a flu, spread quickly.

The mystery illness seemed like pneumonia, causing muscle aches, coughs, diarrhea, chest pains, high fevers, and headaches. As public health officials investigated, it became apparent that the illness was not contagious. Antibiotics were also effectively treating the hospitalized patients.

The Centers for Disease Control and Prevention (CDC) joined state officials in investigating this illness. Fears of biowarfare took hold, and epidemiologists examined not only the sick members but also the venue in which they all became ill. The media is actually responsible for settling on the name "Legionnaires' disease" after the American Legion members afflicted.

Months and months after the outbreak, the CDC finally solved the mystery. A previously undiscovered bacterium was present in the tissue from a patient's lung. Microbiologists were able to find that the bacteria thrived in hot weather and in water. During the convention, the bacteria, now named *Legionella*, were able to spread through the air conditioning system of the hotel.

With the discovery of the *Legionella* bacterium, the importance of maintaining water supplies and air conditioning systems, shown here, to keep them bacteria free was stressed.

bacterial infection that affects a majority of people, particularly during the teenage years. Oil clogs the tiny pores of the skin. Bacteria collect and feed on the oil, multiplying until an acne sore results. An increase in the skin's oil production during the teen years makes young people particularly susceptible to acne.

Caused by the *Helicobacter pylori* bacterium, an ulcer is a sore in the lining of the stomach or duodenum, a part of the small intestine. Often called peptic ulcers, these infections can produce symptoms of abdominal pain, nausea, vomiting, bloating, and heartburn.

Commonly called pink eye, conjunctivitis is an inflammation of transparent membranes in the eyeball and eyelid. Staphylococci, pneumococci, and *Haemophilus influenzae* bacteria cause most cases of bacterial pink eye, which is highly contagious. Conjunctivitis can also result from allergies, viruses, and other causes.

Like other germs, bacteria can make us ill in a variety of ways, but we aren't defenseless. Our bodies are prepared to fight back, and when they can't fight on their own, science and medicine can often step in to help.

WHAT DO YOU THINK?

1. Can you name some other foods not mentioned in this chapter that might be good at resisting bacterial growth?

2. Why do you think bacteria such as *Legionella* are still being discovered in modern times?

3. How can one condition, such as pink eye, be caused by different germs?

Proper handwashing is our most basic defense against germs. Cleaning with soap and water is effective in removing bacteria from our skin, even without special antibacterial agents.

CHAPTER 3

HOW DO WE FIGHT BACTERIA?

The fight against germs is a never-ending war for humankind. However, as our understanding of microbes has continued to increase, the fight has become fairer. First and foremost, like other germs, bacteria can be thwarted with basic hygiene. Proper handwashing can remove most germs from your skin and prevent them from entering the body through the eyes, nose, or mouth. Brushing your teeth can clear bacteria, which love the mouth's moisture and warmth, and keep your gums, teeth, and mouth healthy. Besides these basic measures that we teach children from a young age, we have other defenses too.

THE IMMUNE SYSTEM

The human body has natural defenses against the entry of harmful bacteria. On the skin's surface, harmless bacteria live, protecting their turf against invading bacteria. Mucous membranes and tiny hairs in the nose trap bacteria, where they are expelled when the nose is blown. Bacteria entering through the mouth or nose may also be killed by mucous membranes in the mouth or throat. Those making their way to the stomach are killed by acids there. Tears in the eyes and saliva in the mouth form additional defenses.

Think about the last time you got something in your eye. You probably teared up. These tears don't happen because of emotions but instead as a part of your first line of defense. Your eyes tear up to try to remove foreign organisms before they make their way into the body.

Cuts, scrapes, skin punctures, and animal bites provide pathways for bacteria to enter the body. However, even these temporary entryways are protected. When the skin is cut, for example, some of the damaged cells release histamine. This chemical signals the body to send more blood to the injury, and the area swells. The blood clots, sealing the cut. Besides that, white blood cells rush to the injury. Called phagocytes, they eat invading bacteria. Afterward, the phagocytes die, and the mass of dead phagocytes and demolished bacteria remains as pus. Other immune system white cells called lymphocytes produce chemicals to attack and kill bacteria and other invaders. Specific lymphocytes target specific kinds of bacteria.

Despite the body's defenses, some pathogenic microbes find a way in and survive. Once inside the body, the bacteria carry on with the business of living—multiplying, eating, and excreting waste. Infections result when they multiply to numbers great enough to interfere with bodily functions or when their poisonous waste products—toxins—cause damage.

Food Safety

Pasteurization is a method of killing harmful bacteria in some fresh products such as milk and fruit juices. For example, milk and cream bottled for sale to the public is pasteurized. One method of pasteurizing fresh milk is to heat it for 30 minutes at 145 degrees Fahrenheit (63 degrees Celsius). A quicker method of killing the bacteria is to heat it to 161° F (72° C) for 15 seconds. Immediately following this, the milk is cooled rapidly to preserve its flavor.

Special care must be taken when canning foods. In the canning process, foods are heated to kill bacteria and then sealed in

When we purchase milk from a grocery store, we can be sure harmful bacteria have been killed during pasteurization.

sterile jars or cans. Thus prepared, foods can be stored without refrigeration or freezing for many months. However, some spores and heat-tolerant bacteria called thermophilic bacteria survive high temperatures. If sufficient heat is not used for the proper length of time, the microbes survive. They grow in the canned food or become active when the container is opened.

Keeping hands washed and clean can also keep bacteria out of food that is being served and prevent cross contamination between foods. For example, *Salmonella*, which is often found on raw chicken, may be killed when the chicken is cooked to the proper temperature. However, if after handling the raw chicken, the cook touched other food without washing their hands, that bacteria might end up on a food that won't undergo heat or cooking, such as a salad.

Handling raw chicken increases the risk of exposure to *Salmonella* bacteria. When handling raw chicken, be sure to thoroughly clean anything that touches it, including utensils, surfaces such as counters and cutting boards, and your hands, before preparing or handling other food.

Vaccines, Antibiotics, and Antitoxins

The human body produces natural defenses, called antibodies, to fight disease-causing bacteria. To combat the microorganisms' toxins, the body produces antitoxins. Vaccines and antibiotics are manufactured weapons used to prevent or fight infections. These human-made forms of treatment are prepared using bacteria or their poisonous waste. Vaccines and antitoxin injections are preventive, meaning they are given when no infection is present in order to protect against a future attack. Antibiotics are used to treat infections that have already set in.

Disease-producing bacteria are used to create vaccines that prevent infections. Given as an injection, a vaccine contains dead or weakened bacteria. The dead bacteria stimulate the human body to react as it would to live bacteria—by producing antibodies. These antibodies protect the individual against the specific kind of bacteria contained in the vaccine. Vaccines generally protect people for a year or longer.

Scientists use bacterial waste products to make antitoxin injections. To make these solutions, the scientist first injects toxins into living animals. An animal's body then produces antitoxins in response to the injected toxins. Next, the scientist draws blood from the animal and uses it to make an antitoxin for use in humans. Antitoxins work more quickly than vaccines, but they don't last as long—only a few weeks or months.

Bacterial infection is often treated using drugs called antibiotics. These are made up of a secretion from some types of bacteria and other microorganisms, which in turn fights bacterial infectios. Bacitracin and polymyxin are two well-known antibiotics. Streptomyces, mold-like substances that exist in soil, are a group of bacteria that are utilized to create tetracycline, neomycin, and many other different classes of antibiotics. Among the best known and effective antibiotics is penicillin, which is derived from the mold known as *Penicillium notatum*.

Along with vaccines, antibiotics are among the most valuable weapons against pathogens and toxins that break through the body's natural defenses. Nations with well-developed health care systems have access to great amounts of antiobiotics and have subsequently eliminated many types of infections among their populations.

One issue that has arisen in recent years, however, has been the growth of bacterial antibiotic resistance. When antibiotics are widely

The antibiotic penicillin is derived from the mold *Penicillium notatum*. Alexander Fleming discovered it by accident when he returned from a vacation and found mold growing on a culture plate contaminated by the bacterium *Staphylococcus*.

used, the strains of bacteria that resist them will tend to survive more than weaker strains, an **evolutionary** process that many organisms undergo and that is often referred to as natural selection.

As these stronger strains grow in population, the antibiotics will be less and less effective in treating these bacterial infections. This is why many health professionals recommend not resorting to antibiotics for sicknesses such as the common cold or the flu, which are caused by viruses.

Our bodies are made with a built-in defense system to keep us safe and healthy. As scientists continue to understand more about how bacterial infections affect humans, we can take steps to reduce the likelihood that they make us sick, and we can be ready when the steps our bodies naturally take aren't quite enough.

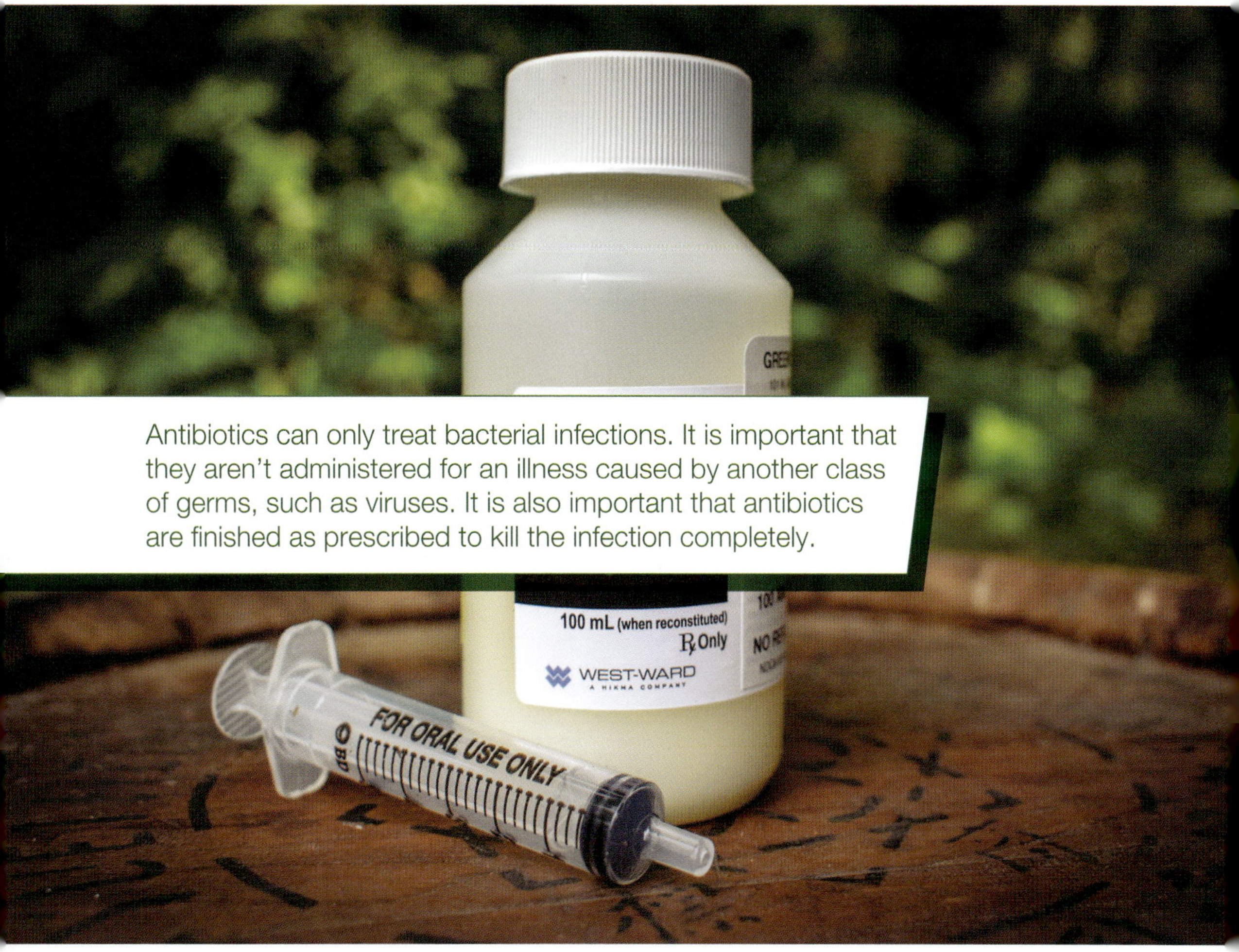

Antibiotics can only treat bacterial infections. It is important that they aren't administered for an illness caused by another class of germs, such as viruses. It is also important that antibiotics are finished as prescribed to kill the infection completely.

ANTIBIOTIC-RESISTANT BACTERIA

Methicillin-resistant *Staphylococcus aureus*, or MRSA, is one of the most problematic antibiotic-resistant bacteria. Anyone can get MRSA. In fact, about 5 percent of patients hospitals carry MRSA in their nose or on their skin. In the community, it spreads by direct contact with infected people or things. This could be as obvious as contact with a wound that contains MRSA bacteria, or something as simple as sharing a towel.

Carrying the MRSA bacterium doesn't mean you have a MRSA infection. Several risk factors increase likelihood of infection, and basic hygiene and wound care can prevent it. However, once a MRSA infection occurs, it can become severe. It can cause bloodstream infections, pneumonia, and surgical site infections in health care settings. Without treatment, it can cause sepsis, which is an extreme and potentially deadly response to an infection in the body.

MRSA is a serious threat because it is difficult to treat. Several antibiotics no longer work to help rid the body of the bacteria. Over time, the bacteria have adapted to become resistant to these drugs. While preventing MRSA infections and working on the development of new antibiotics to treat it are important steps, reducing the use of unnecessary antibiotics is a major factor in preventing future bacteria from becoming resistant too.

WHAT DO YOU THINK?

1. How does our immune system know which bacteria to fight?
2. Why are vaccines such an important part of the fight against germs?
3. Why is it important to take antibiotics as prescribed, even after you feel better?

Instead of spraying chemical pesticides, some crop-dusting planes are able to spread bacteria to keep pests at bay on crops.

CHAPTER 4

WHEN ARE BACTERIA HELPFUL?

Not all bacteria are germy microbes that make us sick. We actually rely on many types of helpful bacteria in our everyday lives. Even some of the pathogenic bacteria have been studied and used to our advantage over time.

STUDY OF BACTERIA

Drug makers are always on the lookout for bacteria that may prove useful in fighting diseases, sometimes known in recent years as **biopharmaceuticals**. In pharmaceutical laboratories, scientists search for microbes to use in new drugs. Much like miners prospecting for gold, these scientists screen large numbers of microbes in order to locate useful ones. Many lifesaving new drugs result from this tedious work.

Many kinds of bacteria develop resistance or immunity to antibiotics that previously killed them. Bacteria's ability to adapt drives scientists to search for newer drugs that will be effective, at least for a time. Viruses known as **bacteriophages** attack and kill specific bacteria. Scientists are turning to viruses in search of new means of combating bacterial infections. Some *E. coli* bacteria, for example, fall victim to certain viruses.

The study of bacteria in laboratories results in more than just drugs to fight infections. Bacteria are used to rid crops of pests naturally. Natural pesticides called bioinsecticides repel pests or kill them outright. For instance, the *Bacillus thuringiensis* (Bt) bacterium kills pests on crops and mosquitoes. Harmless to humans, it is a safe alternative to chemical pesticide and is used worldwide. Some bioinsecticides are mixed with cornstarch in a special process that prepares them to be spread on crops by crop-dusting aircraft.

INDUSTRIAL USES

Besides being important to laboratory research and farming, bacteria are also used in mining. Called biomining, a process using the *Thiobacillus ferrooxidans* bacterium extracts metals from ore. This microbe gets its energy from combining oxygen with inorganic material such as minerals. In doing so, the bacterium releases acid that washes metals out of ore. Miners remove copper from ore using ferrooxidans bacteria. Gold, too, is separated from ore using biomining.

Thanks to genetic engineering, a kind of bacteria has been developed for the purpose of eating oil. Oil spills such as the Exxon Valdez spill in 1989 coat huge areas of water with oil. As a result, vast numbers of fish, birds, and other wildlife suffer or die. The 11 million-gallon (41.6 million-liter) Exxon Valdez accident created an oil slick covering nearly 100 square miles (260 sq km) off the coast of Alaska.

For some bacteria, this oil is a feast. In the early 1970s, scientists genetically created bacteria that could eat oil. To clean up an oil slick, sawdust is sprinkled over it. The oil soaks into the sawdust, forming globs that sink to the bottom. Here bacteria devour the oil,

breaking it down and making it harmless. Another method of cleaning an oil slick is sprinkling it with clay. Like the sawdust, the clay absorbs the oil, only it does not sink. Bacteria then break down these clumps of oil-soaked clay.

Oil spills are devastating to the environment and the animals and plants in it. Because of research with bacteria, we now have more tools to help clean up the mess when it happens.

Good Bacteria in Food

The changes brought by bacteria in food can be a good thing. **Fermentation** is the key in the production of many foodstuffs. It is a chemical process brought about by certain microorganisms, including some kinds of bacteria. For example, *Lactobacillus bulgaricus* bacteria change milk sugar into lactic acid, a process that produces buttermilk, cream, and yogurt. So-called live yogurt contains living lactobacilli. A seemingly endless variety of cheeses are made using different combinations of bacteria, each blend resulting in a different flavor. Other foods and beverages made with the help of bacteria include pickles, soy sauce, sauerkraut, wine, beer, coffee, chocolate, and some cured meats, among others.

Dill pickles, for example, are the end product of fermenting cucumbers. Streptococci bacteria begin the fermentation process, and as the level of acidity of the cucumbers falls, the leuconostoc and pediococcus bacteria species continue the process, helped along by *Lactobacillus plantarum* as well.

Some of these foods help us maintain the helpful bacteria in our body. We house an estimated 100 trillion good bacteria that help our bodies digest food, absorb nutrients, and produce vitamins!

Pickles have bacteria as a key ingredient.

Bacteria are one of the most numerous microorganisms on Earth. Scientists even say we have as many bacteria and other microbes in our body as cells. Sometimes the diseases bacteria cause can be frightening, but we need to remember how vital bacteria are to not only our own lives, but to the world around us as well. Scientists are driven to study and learn more about these microbes and will certainly make great new discoveries related to them in the future. For organisms that are millions of years old, bacteria still bring new surprises.

TASTY BACTERIA

When you hear the word "bacteria," you likely think of germs that make people sick. You probably don't think of some of your favorite foods. However, if you enjoy yogurt, bacteria are part of your diet! The good bacteria in yogurt are often called probiotics. This term comes from the Latin root *pro* meaning "for" and the Greek word *biotikos* meaning "pertaining to life." Essentially, the idea is that these bacteria are good for your life, and for your gut.

When you are prescribed a strong antibiotic, your pharmacist may suggest you take it along with a food such as yogurt. This is because although the antibiotic is prescribed to kill the bad bacteria making you sick, it also unfortunately can kill some of the good bacteria keeping your gut healthy. This can lead to the unwanted side effects of diarrhea and upset stomach. Eating a food such as yogurt helps replenish those good bacteria or probiotics before you can feel unwell from losing them.

Besides being good for you, the bacteria in yogurt make it what it is. The bacteria ferment the dairy used to make the yogurt, just as they do for cheese. So the next time you take a big spoonful of yogurt, you can thank those tasty bacteria!

WHAT DO YOU THINK?

1. Why do scientists continue to study bacteria that we already have treatments for?
2. Why are bacteria a good way to keep crops free of pests?
3. How can you tell the difference between good and bad bacteria in food?

GLOSSARY

anaerobe: An organism that lives in the absence of oxygen.

bacteriophage: A virus that attacks and kills bacteria.

biopharmaceutical: A medicinal product made from or extracted from biological sources, such as bacteria.

carcass: The dead body of an animal.

conjugation: A method of sexual reproduction used by some kinds of bacteria in which genetic information is exchanged.

enzyme: A substance produced by a living organism that helps bring about a specific biochemical reaction.

evolutionary: Relating to the gradual development of something, often in reference to living organisms.

fermentation: The chemical breakdown of a substance by bacteria, yeasts, or other microorganisms.

fission: Also called binary fission, the reproduction of cells through division. One cell divides into two identical clones.

flagellum: (plural **flagella**) A long, whiplike tail on some microscopic creatures, including some bacteria. Flagella provide a means of movement.

microbe: A microscopic organism, including those that can cause disease.

organic: Relating to or coming from living material.

pasteurization: A process whereby liquids or foods are heated to specific temperatures for certain lengths of time to kill microscopic organisms.

prokaryote: A microscopic, single-celled organism that does not have a distinct nucleus with a membrane or other specialized organelles.

FIND OUT MORE

BOOKS

Arbuthnott, Gill. *Microbe Wars*. Dorking, UK: Templar Publishing, 2021.

Brundle, Joanna. *Antibiotics*. New York, NY: Enslow Publishing, 2020.

Mayer, Melissa. *The Micro World of Viruses and Bacteria*. North Mankato, MN: Capstone Press, 2022.

WEBSITES

American Museum of Natural History: Microbiology
amnh.org/explore/ology/microbiology
This section of the American Museum of Natural History website explores microbiology—the study of microscopic living things.

Kids Discover: Bacteria
online.kidsdiscover.com/unit/bacteria
This unit on the Kids Discover website covers bacteria and microbiomes with vocabulary activities.

National Geographic Society: Bacteria
nationalgeographic.org/encyclopedia/bacteria/
This encyclopedic entry on bacteria also includes links to related resources on the topic.

ORGANIZATIONS

American Public Health Association (APHA)
800 I Street NW
Washington, D.C. 20001-3710
Email: comments@apha.org
www.apha.org
The APHA is a professional organization for public health practitioners in the United States, championing the health of all people and communities.

Association of Medical Microbiology and Infectious Disease Canada (AMMI)
192 Bank Street
Ottawa, ON K2P 1W8
Canada
Email: info@ammi.ca
www.ammi.ca
The AMMI promotes the prevention, diagnosis, and treatment of human infectious diseases through its involvement in education, research, clinical practice, and patient advocacy.

Centers for Disease Control and Prevention (CDC)
1600 Clifton Road
Atlanta, GA 30329
www.cdc.gov
The CDC is the primary U.S. government agency in charge of tracking, treating, and preventing public health threats, including infectious diseases.

National Institutes of Health (NIH)
9000 Rockville Pike
Bethesda, MD 20892
Email: NIHinfo@od.nih.gov
www.nih.gov
The NIH is a research facility in the Washington, D.C., area that is the primary agency of the U.S. government in charge of biomedical and health-related research.

INDEX

About the Author

Kathryn Kroe has a master's degree in public health from the University at Buffalo and a bachelor's degree in education from SUNY Buffalo State College. She is passionate about spreading accurate information and helping young people understand issues pertaining to their health and the health of those around them. She currently lives in Lockport, New York, with her husband and two toddlers. She enjoys walking with her family, playing with her Australian shepherd, Zoey, and traveling to see new places.